Edwin J. Houston

Algebra Made Easy

Being a clear explanation of the mathematical formulae found in Prof.

Thompson's Dynamo-electric machinery and Polyphase electric currents

Edwin J. Houston

Algebra Made Easy
Being a clear explanation of the mathematical formulae found in Prof. Thompson's Dynamo-electric machinery and Polyphase electric currents

ISBN/EAN: 9783337393915

Printed in Europe, USA, Canada, Australia, Japan

Cover: Foto ©berggeist007 / pixelio.de

More available books at **www.hansebooks.com**

ALGEBRA MADE EASY

Being a clear explanation of the Mathematical
Formulæ found in Prof. Thompson's
Dynamo-Electric Machinery and
Polyphase Electric Currents

BY

EDWIN J. HOUSTON, Ph. D.,

AND

ARTHUR E. KENNELLY, Sc. D.

———

NEW YORK

AMERICAN TECHNICAL BOOK CO.

45 VESEY STREET

1898

PREFACE.

This little volume has been prepared by the authors for the purpose of elucidating the mathematical formulæ appearing in the pages of Professor Silvanus P. Thompson's "Dynamo-Electric Machinery," and "Polyphase Electric Currents." While the authors do not profess to make their readers mathematicians, they believe that their readers will be fully capable of understanding the meaning of all the mathematical formulæ appearing in the above-mentioned works, after a careful perusal of this little book.

PHILADELPHIA, *December*, 1897.

CONTENTS.

ALGEBRA MADE EASY.

CHAPTER I.

INTRODUCTION.

THE essential difference between pure and applied mathematics lies in the fact that symbols are employed in pure mathematics for the purpose of conveniently studying the relations between the quantities they represent, entirely independently of arithmetical or practical applications; whereas, in applied mathematics the symbols are employed especially for the purpose of enabling practical and arithmetical solutions and applications to be obtained from the expressions of the laws controlling such quantities.

Just as there is no limit to infinite truth,

so there is no limit to the range, extent, and complexity of pure mathematics; but applied mathematics is limited in range, in order to be capable of ready application and utilization. When a formula or analysis in the department of applied mathematics becomes so complex, difficult, or intricate, as to render its solution and arithmetical computation more laborious than the object to be attained deserves, it thereby places itself beyond the pale of applied mathematics. Consequently, applied mathematics is relatively simple mathematics.

The mathematics which the engineer employs must be relatively simple, because his duties compel him to adopt methods of computation that shall be readily susceptible of being checked and corroborated, and shall not be so intricate as to demand undue share of his time and thought. Anyone who can master arithmetic can master all the processes of applied mathematics, such as the engineer has to use, since such mathematics has to be thought out and worked out in arithmetic.

There never has been, and, proverbially, there never can be, a royal road to knowledge, the pathway to which is only found on the highway of labor. It is neither the intention nor the claim of the authors, in the following pages, to make their readers competent mathematicians. But it is their intention and claim to make them able to grasp and understand the meaning of the formulæ and equations which are scattered throughout technological literature. This symbolic language, which so largely pervades scientific technology, is the natural and beautiful language of exact quantitative expression. It is essentially a simple language, shorn, by long and wearisome evolution, of almost every vestige of unnecessary or superfluous appendage, and which, when properly enunciated, carries a meaning to the student as clear and perspicuous as its expression is brief and direct.

To handle and manipulate algebraic expressions, to solve equations and reduce them to their simplest forms, is an art attained only by study and practice, and with which the following pages do not deal. It has no essential part in the under· standing of mathematical expressions.

CHAPTER II.

+ (Plus). The sign of addition. As 7 + 5, meaning the sum of five and seven; *i. e.*, seven added to five, or five added to seven.

= (Equality). The sign of equality. As 7 + 5 = 12; meaning that the sum of seven and five *is equal* to twelve.

− (Minus). The sign of subtraction. As 7 − 5 = 2 ; meaning that five subtracted from seven is equal to 2.

× (Multiplication). The sign of multiplication. As 7 × 5 = 35 ; meaning that 7 multiplied by 5, or 5 multiplied by 7, is equal to 35.

÷ (Division). The sign of division.

As $7 \div 5 = 1.4$; meaning that seven divided by 5 is equal to $1\frac{4}{10}$.

$\therefore$ (Therefore). A sign used in mathematical reasoning as a mere symbol for the word " therefore."

$:: :$ (Ratio). Signs of proportion. As $7 : 5 :: 14 : 10$; meaning seven *is to* five *as is* 14 *to* 10.

( ) (Brackets). Various forms of parentheses or brackets, employed for grouping into one mass a compound quantity. Thus $5 \times (7 + 5) = 5 \times 12$.
{ }
[ ]

a, b, c, d, etc., (Symbols). Letters
A, B, C, D, etc.
of the alphabet representing quantities; usually, but not necessarily, known constant quantities. Thus g, is a symbol commonly used to represent the gravitational force which the earth exerts upon a gramme mass. In scientific units $g = 980.07$ dynes at the sea-level in Washington, D. C.; consequently, g, is more than a mere number—it stands for a certain

force having a magnitude of 980.07 units,
the unit being the dyne.

Again, π (Greek letter Pi) is a symbol
commonly used to represent the ratio between the circumference of a circle and
the diameter, so that,
Circumference of a circle = π × Diameter of the circle.

Here π = 3.1416, approximately; or,
roughly, 3⅐. In this case, π, is the symbol
of a mere numerical magnitude, or ratio
between two geometrical quantities.

A symbol, therefore, may stand for a
number considered solely as such; or for
a number representing any particular
quantity, physical, astronomical, chemical,
etc., stated in reference to a particular
unit. The symbols x, y, z, X, Y, Z, are
commonly, but not necessarily, used in the
mathematical statement of relations, for
quantities whose values are unknown, and
which may or may not be determined from
the relationship given.

A formula is a rule mathematically expressed, for determining the value of any quantity. Thus the equation:

The circumference of a circle $= \pi \times$ Diameter,

is a simple formula from which the circumference of a circle becomes known as soon as its diameter is given.

ADDITION.

In the equation

$$c = a + b,$$

we have a symbolic form for the following statement:

c, is equal to the sum of a and b.

If a, b and c, are mere numbers, and $a = 5$, while $b = 7$; then $c = 12$, because $12 = 5 + 7$.

If a and b, are symbols which represent electromotive forces acting in a circuit, or weights lying in a scale pan, then c, is either a total electromotive force, or a total weight, accordingly. Consequently, a simple equation involving the process of addition may express a mere relation between ordinary numbers, or between numbers which represent physical quantities expressed in terms of units. Thus, the equation,

$$x = a + b + c + d,$$

where $a = 1$, $b = 2\frac{1}{2}$, $c = 3\frac{1}{2}$, $d = 4$, gives the relation $x = 11$; or, the unknown quantity x, in this equation, is known to be equal to 11, because the sum of the *terms* a, b, c and d, on the right-hand side of the equation, is known.

Again, on page 214, of Thompson's "Dynamo-Electric Machinery," appears the following equation:

$$C_a = C + C_s$$

Where, C_a, termed "C sub a," is the symbol expressing the current strength in the armature of a shunt-wound dynamo; C, is the current strength supplied to the main external circuit; and C_s, termed "C sub s," is the current supplied to the shunt field. Here, the subscripts a and s, are used to distinguish between the current strengths in the different portions of the circuit, and the equation makes in symbolic form the following statement:

The current in the armature of the machine is equal to the sum of the currents in the main circuit and the shunt field; or, if $C = 100$ amperes, and C_s, the shunt-field current, is one ampere, then the armature current $C_a = 100 + 1 = 101$ amperes.

Again, on page 188, of Thompson's "Polyphase Electric Currents," occurs the equation:

$$V_{pq} + V_{qr} + V_{rp} = 0.$$

This equation has reference to Fig. 130, on the preceding page. The equation is a symbolic method of concisely stating the following:

The sum of the three electromotive forces which occur in the three branches pq, qr, and rp, represented symbolically by the symbols V_{pq}, V_{qr}, and V_{rp}, is always equal to zero. Consequently, if any pair of these electromotive

forces is, say 50 volts, then the remaining E. M. F. must be equal to − 50 volts;
or, if any particular E. M. F. is, say
25 volts, then the sum of the remaining
pair of E. M. F.'s must be equal to − 25
volts. In other words, when the three E.
M. F.'s expressed in volts, or other units,
are added together, the sum total is zero.

Similar considerations apply to any number of added simple terms, such as are
found in the equation:

$$x = a + b + c + d + e.$$

SUBTRACTION.

As already pointed out, the minus sign
prefixed to a quantity indicates that the
quantity has to be taken negatively, and
has to be subtracted from the quantity
with which it is associated.

Thus in the equation:

$$x = a - b,$$

the right-hand side contains two terms,
the first a, which is + or positive, and the

second $-$ b, which is negative; b, must, therefore, be subtracted from a, or a must have its value diminished to the extent of the value of b. It may happen that b, is greater than a; as, for example, if $a = 3$ and $b = 5$; but the result is interpreted by giving to the difference a negative sign; or, in this case, $x = -2$. Consequently, in an equation containing simple terms, some of which are positive and some negative, it suffices to add all the positive terms for a positive sum, and all the negative terms for a negative sum, and then subtract the latter from the former. Thus:

$$x = 5 + 7 - 3 + 10 - 2,$$
$$\text{or, } x = 5 + 7 + 10 - 3 - 2,$$
$$\text{or, } x = 22 - 5,$$
$$\text{or, } x = 17.$$

On page 510, of Thompson's "Dynamo-Electric Machinery" appears the equation,

$$C_a = C - C_e.$$

Here C, represents the total current

---

supplied to a shunt motor; C_a the current supplied to the armature; and C_s, the current supplied to the shunt field. The equation is equivalent to the following statement:

The current supplied to the armature is equal to the current supplied to the machine, less the current supplied to the shunt field; so that if the total current C, is 100 amperes, and the shunt-field current C_s, 1 ampere; then the armature current C_a, would be $100 - 1 = 99$ amperes.

BRACKETS OR PARENTHESES.

It is often desirable, in expressing an equation, to separate some of the terms into groups by placing them within a bracket. Thus, the equation

$$x = a - b + c + d - e,$$

states that x, is the sum of all the quantities on the right-hand side, after due allowance has been made for their sign;

i. e., after the proper additions and subtractions have been effected. It is often convenient to separate the positive quantities from the negative quantities. The first step is to bring them together in two groups; thus,

$$x = a + c + d - b - e.$$

These groups may be included in brackets to give each the appearance of a single term. Thus,

$$x = (a + c + d) + (- b - e).$$

Here the compound term $+ (- b - e)$ may be written $- (b + e)$; because adding the sum of two negative quantities is the same as subtracting their positive sum, so that the sum of b and e, is to be subtracted from the first compound term $(a + c + d)$. Consequently, a negative sign before a bracket or parenthesis reverses the sign of all the terms within it. Thus, in the above equation let $a = 1$, $b = 2$, $c = 3$, $d = 4$, and $e = 5$.

Then $x = 1 - 2 + 3 + 4 - 5,$
$$= (1 + 3 + 4) + (- 2 - 5),$$
$$= (1 + 3 + 4) - (2 + 5),$$
$$= 8 - 7,$$
$$= 1.$$

The equation might also be written

$$x = \overline{1 + 3 + 4} - 2 - 5.$$

Where the line serves as a pair of parentheses to group 1, 3, and 4, into a compound term,

$$\text{or, } x = [1 + 3 + 4] - [2 + 5],$$
$$\text{or, } x = \left\{ 1 + 3 + 4 \right\} - \left\{ 2 + 5 \right\}$$

That is to say, any form of bracket or parenthesis might be used to separate the two groups of terms from each other.

MULTIPLICATION.

An equation,

$$x = a \times b,$$

means that the quantity x, is the product of a and b; so that if $a = 5$ and $b = 10$, $x = 50$. Where simple terms are employed, as in this case, the multiplication sign may be omitted, and the equation is written,

$$x = a\,b,$$

meaning that x, is equal to the product of a and b. In general, when two symbols follow each other without any sign between them, their product is thus indicated; for example,

$$x = abcd.$$

Here x, is the product of a, multiplied by b, multiplied by c, and multiplied by d. If $a = 1$, $b = 2$, $c = 3$, and $d = 4$; $x = 24$.

In some cases a point or period takes the place of the multiplication sign. Thus,

$$x = a.b.c.d.$$

On page 168, of Thompson's "Dynamo. Electric Machinery," occurs the equation :

$$\text{(average) } E = n\,Z\,N,$$

where E, is the average E. M. F. generated

by a dynamo-electric machine; n, is the number of revolutions of its armature per second ; Z, is a certain number of conductors on the armature surface ; and N, the total number of magnetic lines from one field pole that traverse the armature. Consequently, the equation is equivalent to the following statement:

The average E. M. F. is the product of the number of turns made by the armature per second, a certain number of conductors lying upon the surface of the armature, and the total number of magnetic lines traversing the armature.

Compound terms formed of the products of simple terms or factors may be subjected to addition or subtraction like simple terms.

Thus, on page 189, of Thompson's "Polyphase Electric Currents," occurs the equation :

$$W = V_{mn}b + V_{mo}c,$$

where W, is the power in watts supplied by a triphase star-wound armature, as rep-

resented in Fig. 54, on page 46 ; V_{mn}, is
the effective pressure between the terminals m and n; and V_{mo}, the pressure
between the terminals m and o; the current b, being in the branch n, and the current c, in the branch o. The equation is
equivalent to the following statement :

The total electric power supplied by a
machine is equal to the sum of two compound terms, the first of which is the
product of the current b, and pressure
V_{mn}, and the second, the product of the
current c, and the pressure V_{mo}.

In some cases a factor appears outside
a bracket. For example, on page 189,
of Thompson's " Dynamo-Electric Machinery," appears the equation :

$$E = (R + r_a + r_m)\, C.$$

This is given in relation to series dyna-
mos. E, is the E. M. F. generated by the
machine ; R, the external resistance of the
circuit ; r_a, the resistance of the armature ;
and r_m, the resistance of the magnets as

shown in Fig. 125. C, is the current strength through the circuit in .amperes. This equation expresses the fact that the product of the current strength in the circuit in amperes multiplied by a compound term, which is the sum of all the resistances in the circuit, is equal to the total E. M. F. in the circuit. This is only another way of stating Ohm's law for the circuit.

Since an equation $x = ab$, is the same as $x = ba$; or, since the order of factors in an equation is indifferent, the equation on page 189, of Thompson's "Dynamo-Electric Machinery," may be written:

$$E = C\,(R + r_a + r_m,)$$

or

$$E = CR + Cr_a + Cr_m.$$

This shows that a factor placed outside a bracket is to be multiplied by each term within the bracket. Suppose, for example, that the current strength C, in the circuit, is 100 amperes; that the external resistance

R, is 1 ohm; that the armature resistance r_a, is $\frac{1}{100}$th ohm; and the field magnet resistance r_m, is $\frac{2}{100}$th ohm—then the equation may be written:

$$E = 100 \ (1 + 0.01 + 0.02),$$
$$E = 100 \ (1.03) = 103 \text{ volts};$$

or, multiplying each term within the bracket by the factor outside,

$$E = 100 + 1 + 2 = 103 \text{ volts.}$$

DIVISION.

The operation of division is represented in algebra either by the *division bar* ÷, or, more commonly, by the *fraction bar* employed in ordinary arithmetic. Thus

$$x = a \div b,$$

may be written

$$x = \frac{a}{b}.$$

If $a = 3$ and $b = 4$, this becomes $x = 3 \div 4$, or $x = \frac{3}{4} = 0.75$. Sometimes the bar takes an inclined form, when it is called a *solidus*, thus, $x = a/b$.

Take, for example, the expression for Ohm's law, which is written in the International system of notation

$$I = \frac{E}{R}.$$

If E, the E. M. F., is 100 volts, and R, the resistance, is 5 ohms, the current strength $I = \dfrac{100}{5} = 20$ amperes, and the equation is equivalent in this case to the following statement:

The current strength in amperes, in a circuit having an E. M. F. of 100 volts, and a resistance of 5 ohms, is equal to 100 divided by 5, or to 20 amperes.

Thus, on page 341, of Thompson's "Dynamo-Electric Machinery," occurs the equation which is referred to as the fundamental equation of the continuous-current dynamo:

$$E = nZN \div 10^8,$$

where n, is the number of revolutions of

the armature per second, Z, is the number of armature conductors employed, N,
is the number of magnetic lines through
the armature, and 10^8, is 10 raised to the
8th power, or 1 followed by 8 zeros. The
equation may be written:

$$E = \frac{nZN}{10^8} \text{ or } \frac{nZN}{100,000,000} \text{ volts.}$$

This is solved by multiplying together
the proper numerical values of n, Z, and N,
and dividing the product by 100,000,000.
The quotient is the E. M. F. of the machine in volts.

On page 493, of Thompson's "Dynamo-
Electric Machinery," there occur the equations:

$$w = EC = E \frac{(\mathcal{E} - E)}{R},$$

where w, is the power utilized in a motor,
E, is the C. E. M. F. of the motor,
C, is the current through the armature; R,
is the resistance of the motor, and $\mathcal{E}$, is
the pressure at the motor terminals. This

is an expression which practically contains
three equations; namely,

$$w = EC \qquad (1)$$

$$EC = E\frac{(\mathcal{E} - E)}{R} \qquad (2)$$

$$w = E\frac{(\mathcal{E} - E)}{R} \qquad (3)$$

These equations may be interpreted as
follows :

(1) The useful power in watts is equal
to the product of the C. E. M. F. of the
motor in volts, and the current strength
passing through its armature in amperes.

(2) The product of the C. E. M. F. in
volts and current in amperes in the armature, is equal to the product of the C. E.
M. F., multiplied by the difference between
the pressure in volts at terminals and the
C. E. M. F. in volts, divided by the resistance of the machine in ohms.

(3) The useful power of the motor in
watts is equal to the last-named quantity.

Thus, if $\mathcal{E}$, the pressure at terminals, $=$

120 volts, $R = 0.05$ ohm, $C = 100$ amperes, $E = 115$ volts; then, by (3),

$$w = 115 \, \frac{(120 - 115)}{0.05}$$

$$= 115 \, \frac{(\, 5 \,)}{0.05}$$

$$= 115 \times 100$$

$$= 11,500 \text{ watts.}$$

CHAPTER III.

IF we multiply a number by itself, the number is said to be *squared*, and is represented by an exponent, index, or superscript 2, placed close to and above the number. Thus,

$$a \times a = a^2; \text{ or, } 10 \times 10 = 10^2 = 100.$$

Similarly, if a number be multiplied by itself and again by itself, it is said to be *cubed*. Thus

$$a \times a \times a = a^3$$

Similarly,

$$a \times a \times a \times a = a^4, \text{ and so on,}$$

This general rule is expressed symbolically by

$$a \times a \ldots \text{ to } n \text{ terms} = a^n;$$

25

where n, is any whole number or integer. Thus,

$$10^8 = 100,000,000,$$

or 1 followed by 8 zeros.

Thus on page 493, of Thompson's "Dynamo-Electric Machinery," occurs the equation:

$$W = w + C^2R.$$

Where W, is the power in watts supplied to the motor; w, is the power in watts utilized in the motor; C, the current strength in amperes passing through the motor; and R, the resistance of the motor in ohms. This means that the total power supplied is equal to the sum of two terms, one being the utilized power, and the other the product of the square of the current and the resistance of the motor. Thus, if $w = 10$ kilowatts, or 10,000 watts; $C = 50$ amperes, $R = 0.1$ ohm, then,

$$W = 10,000 + 50 \times 50 \times 0.1 = 10,250$$

watts.

On page 144, of Thompson's " Polyphase Electric Currents," there is an equation :

$$T = q \ \frac{sR}{R^2 + k^2s^2}.$$

Here T, which is the torque of the motor, is expressed as the product of q, and a fraction. The numerator of the fraction is sR, or the product of s and R ; the denominator of the fraction is the sum of two terms : namely, the square of R, and the square of k, multiplied by the square of s. This equation might be written :

$$T = \frac{qsR}{R^2 + k^2s^2},$$

just as 5 (3/4) may be written 15/4.

On page 152, of Thompson's " Polyphase Electric Currents," appears the equation :

The rotor heat $H = K (\Omega - \omega)^2$.

Here Ω (Capital Omega) is an angular velocity of a rotary magnetic field expressed in unit angles per second, and ω, (small Omega) is the angular velocity of the rotor, or revolving member of the motor. Consequently, $(\Omega - \omega)$ is the difference between two angular velocities, or the angular velocity of the field relatively to the moving armature or rotor, so that if the armature revolves at exactly the same speed as the field $(\Omega - \omega) = 0$. Then the equation states that the heat H, is the product of K, and the square of the quantity $(\Omega - \omega)$. K, itself is stated at the top of the page to be:

$$\frac{Z \; x_{m.}^2}{4r}$$

In the same way, any combination of terms may be raised to any power. Thus,

$$(a + b - c + d)^5$$

means that the quantity $(a + b - c + d)$ must be multiplied by itself four times in succession. The first product would be the square, the second product would be the cube, the third product would be the fourth power, and the fourth product would be the fifth power.

If we multiply 10^2 or 100, by 10^3 or 1000, we know that the product is 10^5 or 100,000. Similarly, if we multiply a^2 by a^3, the product $a^2a^3 = a^5$, and not a^6. This rule, which is of general application, shows that when products are formed of the powers of a quantity, the indices or exponents of the powers are added together to form the product.

According to this rule, $10^3 \times 10^0 = 10^{(3+0)}$. Here the index is the sum of 3 and 0, or simply 3, representing 10^3. From this it is evident that if we multiply 10^3 by 10^0, we leave it unaltered, just as it would be if multiplied by unity, because $10^3 \times 1 = 10^3$. Consequently, $10^0 = 1$. This relation is generally true and is expressed by the equation

$$a^0 = 1,$$

or

$$x^0 = 1,$$

always, whatever a or x may be.

Again, $10^3 \times 10^{-3} = 10^{(3-3)} = 10^0 = 1$;

.·. dividing both sides of the equation,

$$10^3 \times 10^{-3} = 1$$
by 10^3,

we obtain

$$\frac{10^3 \times 10^{-3}}{10^3} = \frac{1}{10^3}.$$

Here, on the left-hand side, the 10^3, in the numerator cancels the 10^3, in the denominator; so that we obtain

$$10^{-3} = \frac{1}{10^3}.$$

Similarly,

$$10^{-1} = \frac{1}{10^1},$$

or, is the *reciprocal* of $10^1 = 0.1$.

$$10^{-2} = \frac{1}{10^2},$$

or, is the reciprocal of $10^2 = 0.02$.

$$10^{-8} = \frac{1}{10^8},$$

or, is the reciprocal of $100,000,000. = 0.000,-000,01$, and generally,

$$a^{-n} = \frac{1}{a^n,}$$

or, is the reciprocal of a^n, or of $a.a.a.$ to n times.

RADICALS OR ROOTS.

By the general law of the summation of exponents,

$$a^2 = a^1 \times a^1 = a^{(1+1)}$$
$$a = a^1 = a^{\frac{1}{2}} \times a^{\frac{1}{2}} = a^{(\frac{1}{2}+\frac{1}{2})}.$$

Here the fractional index or exponent, represents what is called a *root*.

For, $a^{\frac{1}{2}}$ is obviously, by the last equation, that quantity which multiplied by itself gives a, or is the *square root* of a. Thus, if $a = 9$, $a^{\frac{1}{2}} = 3$. $a^{\frac{1}{2}}$ is often written $\overset{2}{\sqrt{a}}$ or $\sqrt{a}$. Again,

$$a = a^1 = a^{\frac{1}{3}} \times a^{\frac{1}{3}} \times a^{\frac{1}{3}} = a^{(\frac{1}{3}+\frac{1}{3}+\frac{1}{3})}.$$

Here $a^{\frac{1}{3}}$ is the quantity which, cubed, gives a, or is the *cube root* of a, and may be written $\overset{3}{\sqrt{a}}$. Similarly, the square root of any quantity, such as ab, is written

$\sqrt{ab}$. or $\sqrt{(ab)}$. Thus, on page 194, of Thompson's " Dynamo-Electric Machinery," there is an equation,

$$R = \sqrt{r_a\, r_s}. \qquad \text{[XII]}$$

This means that the resistance R, is equal to the square root of the product of two resistances; one of which is the resistance of the armature, and the other the resistance of the shunt field-magnets. If the product, $r_a\, r_s$, is represented by the symbol r; or, if we assume that $r_a\, r_s = r$, then it would follow that

$$R = \sqrt{r}.$$

The square root of the product of two quantities is called their *geometrical mean*, so that the equation [XII] declares that R, is the geometrical mean of the two resistances, r_a and r_s.

In a similar manner, $r^{\frac{1}{n}}$ is the quantity whose nth power gives r, and may be written $\sqrt[n]{r}$. Thus,

$125^{\frac{1}{3}} = \sqrt[3]{125} = 5$, because $5 \times 5 \times 5 = 125$; or, 5, is the cube root of 125.

We see, therefore, that

$$x^2 = x \times x;$$

or, is the square of x.

$$x^{-2} = \frac{1}{x \times x}$$

the reciprocal of the square of x.

$$x^{\frac{1}{2}} = \sqrt{x};$$

or, is the square root of x.

In a similar manner we may have any fraction for the exponent of x. Thus, we know by the law of summation of indices that

$$x^{\frac{3}{2}} = x^{\frac{1}{2}} \times x^{\frac{1}{2}} \times x^{\frac{1}{2}} \text{ or } (x^{\frac{1}{2}})^3$$

so that $x^{\frac{3}{2}}$, represents the cube of the square root of x, and may be written

$$x^{\frac{3}{2}} = (\sqrt{x})^3$$

For example, if

$$x = 4, x^{\frac{3}{2}} = (\sqrt{4})^3 = 2^3 = 8.$$

Again,
$$x^{\frac{3}{2}} = (x^3)^{\frac{1}{2}} = \sqrt{x^3}.$$
For example, if
$$x = 4, \; 4^{\frac{3}{2}} = \sqrt{4^3} = \sqrt{64} = 8.$$

So that the square root of the cube of x, is equal to the cube of the square root of 2.

In general
$$x^{\frac{m}{n}} = (x^{\frac{1}{n}})^m = (x^m)^{\frac{1}{n}} = \sqrt[n]{x^m} = (\sqrt[n]{x})^m.$$

On page 137, of Thompson's "Dynamo-Electric Machinery," there is the following equation:
$$W = 0.0033 \times 10^{-7} \times n \times B^{1.6}.$$

This expresses the fact that the power W, in watts, expended per cubic centimetre of iron, by hysteresis, is the product of four quantities: The first of these is the numerical constant 0.0033, or $\dfrac{33}{10,000}$; the second is 10^{-7} or $\dfrac{1}{10^7} = \dfrac{1}{10,000,000}$; the third is n, or the number of magnetic cycles executed in the iron per second, while the fourth is $B^{1.6}$, or the $1\frac{6}{10}$th power of B, the magnetic density in the iron, expressed in C. G. S. units per square centimetre.

$B^{1.6} = B^{\frac{16}{10}}$, so that if we form the 16th power of B, and then take the tenth root of this quantity; that is, if we perform the operation B^{16}, and then $\sqrt[10]{B^{16}}$, we obtain the quantity $B^{1.6}$. Or, if we take the tenth root of B, written $\sqrt[10]{B}$, and then take its sixteenth power, we shall obtain the same result. This quantity will obviously be greater than B, itself, or B^{1}, and will be less than $B \times B$, or B^{2}, since 1.6 is inter-mediate between 1 and 2.

On page 160, of Thompson's "Poly-phase Electric Currents," occurs the fol-lowing formula:

$$I_1 = \sqrt{r^2 4\pi^2 (n + m)^2 L^2}.$$

This is equivalent to the following state-ment: The impedance in ohms I_1, is

equal to the square root of a compound
quantity. This compound quantity is
the sum of two terms. The first term
is the square of the resistance in ohms,
while the second term is the product of 4
into the square of the quantity π, which
is 3.1416, approximately, or the ratio of
a circumference to its diameter, into the
square of the sum $(n + m)$ of two frequencies, into the square of the inductance
L, in henrys. This equation might be
written :

$$I_1 = \sqrt{\left\{ r^2 + 4\pi^2 (n + m)^2 L^2 \right\}}$$

or

$$I_1 = \left\{ r^2 + 4\pi^2 (n + m)^2 L^2 \right\}^{\frac{1}{2}}.$$

CHAPTER IV.

EQUATIONS AND THEIR SOLUTION.

THE solution of equations is only to be successfully attained by practice, but the elementary rules for their operation are very simply expressed.

When two things are equal to each other, which is the condition expressed by an equation, the same operation performed upon each will leave the equality unchanged. Thus, if

$$a = b \qquad (1)$$

$$\frac{a}{2} = \frac{b}{2} \qquad (2)$$

$$a^2 = b^2 \qquad (3)$$

$$a + c = b + c \qquad (4)$$

In equation (2) we have divided both sides by 2, and the quotients must remain equal. In equation (3) we have squared both sides of the equation, and the result must be equality. In equation (4) we have added the same quantity c, to both sides of the equation and equality must still subsist.

If $\qquad a + b = c \qquad\qquad$ (5)

then $\qquad a + b - b = c - b \qquad$ (6)

or, $\qquad a = c - b \qquad\qquad$ (7)

From this it is evident that we may carry over a term from one side of an equation to the other by changing its sign; for, in equation (5) the positive quantity b, is on the left-hand side of the equation, whereas in equation (7), which is derived from equation (5) by subtracting b, from both sides, the quantity b, appears on the right hand with the negative sign.

An equation

$$x = a + b + c + d \ldots \text{ etc.,}$$

is an equation of the first degree, because x, appears as of the first power, or x^1.

The equations

$$x^2 = b + c$$

or

$$x^2 + ax = b + c$$

are *quadratic equations*, or *equations of the second degree*, because there occurs a second power of x, in the equation.

Similarly, such an equation as

$$x^3 + 3x^2 + 3x = 5$$

is an *equation of the third degree;* and, generally, an equation involving x^n, is an *equation of the nth degree.*

An equation can always be solved, when of the first, second, or third degree, by definite rules. That is to say, it can be so manipulated, by suitable operations upon both sides, that the value of x, can be obtained. There is, however, no known way of generally solving equations of higher degrees than the third; *i. e.*, equations of the fourth, fifth, sixth, etc., degrees. But the numerical values of the unknown quantity can be obtained by approximation with all desired accuracy by a definite procedure.

Every algebraic expression is referable to the preceding rules; that is to say, to combinations of additions, subtractions, multiplications, divisions, powers, and roots; and, however difficult it may be to solve or manipulate equations, the foregoing explanations will always enable any algebraic equation to be understood, or to be arithmetically solved, when all the symbols are replaced by their proper numerical values. Thus, we may take what is, perhaps, the longest and most complex formula appearing in Thompson's "Polyphase Electric Currents." This appears on page 163, as

$$\text{Torque} = qr\left[\frac{n-m}{r^2 + 4\pi^2\,L^2\,(n-m)^2} - \frac{n+m}{r^2 + 4\pi^2\,L^2\,(n+m)^2}\right]$$

This equation is equivalent to the following statement:

The torque of the motor is the product of three quantities, the first quantity q, is expressed immediately below as

$$q = \frac{ZA^2B^2\pi}{4},$$

that is to say, the quantity q, is one quarter of the product of Z, into the square of A, into the square of B, into 3.1416.

The second quantity is the resistance r.

The third quantity is contained within a pair of brackets and consists of the sum of two fractions.

The first fraction has as numerator the difference between two frequencies n and m, respectively. The denominator of the fraction is the square of r, added to the product of 4, into the square of 3.1416 into the square of the inductance L, into the square of the difference $(n - m)$.*

The second fraction, which is to be subtracted from the first, has, as its numerator, the sum of the two frequencies $n + m$,

* By a misprint, the square of $(n - m)$ has been omitted in the text referred to.

and as its denominator the square of the resistance r, added to the product of 4, into the square of 3.1416, into the square of the inductance L, into the square of the sum of the frequencies $(n + m)$.

CHAPTER V.

LOGARITHMS.

We have seen that by the law of the summation of indices

$$x^2 \times x^3 = x^{2+3} = x^5$$

and similarly

$$x^{1.3} \times x^{3.6} = x^{4.9}$$

If, then, the quantity $x^{1.3} = a$; or, the 1.3d power of the *base* x is a, and $x^{3.6} = b$; or, the 3.6th power of the base x is b, and $x^{4.9} = c$; or, the 4.9th power of the base $x = c$, it follows that $ab = c$.

Suppose that we had a table of indices of a given base, say 5, and that we found from this table that the number 25, was 5^2, or had an index of 2, while the number 125, had the index 3, corresponding to 5^3; then we should know that

$$25 \times 125 = 5^2 \times 5^3 = 5^5,$$

and, if the table informed us that the
number corresponding to the power 5, was
3125, then we should know that:

$$25 \times 125 = 3125,$$

and we should have been saved the
trouble of performing the multiplication.
Here the indices 2, 3, and 5, are the re-
spective *logarithms* of the numbers 25, 125,
and 3125, to the base 5.

The ordinary tables of logarithms are
usually employed for the purpose of enabling multiplication and division to be
effected quickly and conveniently without
actual arithmetical computation. The
base of the ordinary table of *common
logarithms*, as they are called, is 10, so
that, since

$$10^1 = 10 \qquad 10^2 = 100 \qquad 10^3 = 1000$$
$$10^0 = 1$$
$$10^{-1} = 0.1 \qquad 10^{-2} = 0.01 \qquad 10^{-3} = 0.001,$$

it follows that to the base 10, the logarithm of 10 is 1; the logarithm of 100, is

2; of 1000 is 3; of 1 is 0; of 1/10th, or
0.1 is −1; of 0.01 is −2; of 0.001 is −3,
etc. All numbers lying between 10 and
100, will have logarithms lying between
1 and 2. All numbers lying between 100
and 1000, will have logarithms lying between 2 and 3, and so on.

Thus, if we want to multiply 15 by 16;
or have to perform by logarithms the
solution of

$$x = 15 \times 16,$$

we know that the logarithm of 15, lies between 1 and 2, because 15, lies between 10
and 100, and $10^1 = 10$ and $10^2 = 100$.
By reference to a table of seven-
place logarithms, or logarithms carried to
seven decimal places, the logarithm of 15
is .1760913. This is the decimal part or
mantissa. The complete logarithm is
1.1760913, because the characteristic is 1,
and is supplied by the reader. The characteristic distinguishes the logarithm
from that of 0.15, or 0.0015, or 1.5, or 1500,
all of which have the same mantissa, but

differ in their characteristics, their logarithms being respectively $-1 + 0.1760913$,
$-3 + 0.1760913$, 0.1760913 and 3.1760913.

Again, the logarithm of 16, is shown in
the tables to be $.2041200$, and with the
proper characteristic of 1, is written
1.2041200.

We now have

$$15 = 10^{1.1760913}$$
$$16 = 10^{1.2041200}$$
$$\therefore 15 \times 16 = 10^{(1.1760913 + 1.2041200)}$$

or

$$x = 15 \times 16 = 10.^{2.3802113}.$$

Here the number x, has as its logarithm
the number 2.3802113. Its characteristic
is 2, and the corresponding number, therefore, lies between 10^2 or 100, and 10^3, or
1000. The decimal part, or mantissa, is
$.3802113$. This is found in the logarithm
tables to be the logarithm of 240, so that

$$240 = 10^{2.3802113}$$

and

$$x = 15 \times 16 = 240.$$

Again, if we look in the logarithm table for the logarithm of the number 5280, which is the number of feet in a mile, we should find that to seven places of decimals the mantissa is .7226339. This means that $10^{3.7226339} = 5280$. The 3, or *characteristic* of the logarithm, is not given in the table, but is known by the reader, because the number 5280 lies between 1000, for which the logarithm is 3, and 10,000, for which the logarithm is 4, so that he supplies the characteristic when he writes the logarithm down.

Again, if we look for the logarithm of 24,900, which is, approximately, the number of miles around the earth at the equator, we should find the value 4.3961993. Here the characteristic 4, is known because the number falls between 10,000 and 100,000, whose logarithms are 4 and 5 respectively. If now, we add these two logarithms together, we perform in fact the equation:

$$x = 10^{3.7226339} \times 10^{4.3961993} = 10^{8.1188332}.$$

Here the logarithm 8.1188332, consists of the characteristic 8, and the decimal part or mantissa 0.1188332, which in the logarithmic tables corresponds to the number 131472. Since the characteristic is 8, we know that the number x, lies between 10^8 and 10^9, or between 100,000,000 and 1,000,000,000, so that the number is evidently 131,472,000 and is correct as far as 6 places of figures. This product x, is evidently the number of feet around the earth at the equator according to the above calculation. Actual multiplication or arithmetical solution of the equation,

$$x = 5280 \times 24,900 ,$$
gives
$$x = 131,472,000 \text{ feet,}$$

which agrees exactly with the above logarithmic computation.

Again, we know that

$$10^a \div 10^b = \frac{10^a}{10^b} = 10^a \times \frac{1}{10^b} = 10^a \times 10^{-b}$$
$$= 10^{a-b}.$$

Thus, the quotient of two numbers, the dividend of which is 10^a, and the divisor is 10^b, is expressed as 10^{a-b}, so that just as the sum of two logarithms gives the logarithm of their product, the difference of two logarithms gives the logarithm of their quotient.

Thus, if we want to divide 170 by 26, by the aid of logarithms, or solve the equation

$$x = \frac{170}{26},$$

we proceed as follows :

The logarithm of 170, lies between 2 and 3. By tables its mantissa is .2304489. The complete logarithm of 170 is, therefore, 2.2304489.

Similarly the logarithm of 26, lies between 1 and 2. By tables its mantissa is .4149733. The complete logarithm of 26 is, therefore, 1.4149733.

We now have

$$170 = 10^{2.2304489}$$

$$26 = 10^{1.4149733}$$

$$\therefore \frac{170}{26} = 10^{(2.2304489-1.4149733)}$$

$$x = \frac{170}{26} = 10^{0.8154756}.$$

Here the logarithm of the number x, lies between 0 and 1, so that x, is between 1 and 10. The mantissa is .8154756. This is found in tables to correspond to 6.53846. If we divide 26 into 270, by the ordinary arithmetical process, we find in fact that the quotient is 6.53846 as far as 5 decimal places.

For example, on page 156, of Thompson's "Dynamo-Electric Machinery," occurs the equation:

$$\text{Permeance} = 2.274 \times a' \times \log_{10}\frac{d_2'}{d_1'}.$$

This is equivalent to the statement that the permeance is the product of the constant numerical quantity 2.274, into a'', into the logarithm to the base 10; *i. e.*, the common logarithm, of the quotient

$$\frac{d_2}{d_1}.$$

The logarithmic quantity might be considered as the common logarithm of the quotient

$$\frac{d_2''}{d_1''},$$

obtained by first dividing d_2'' by d_1'' arithmetically, and then obtaining the logarithm of the quotient by examining a table of logarithms. But the same result will be obtained if we subtract the logarithm of d_1'' from the logarithm of d_2''. Thus, suppose that $d_2'' = 48$, and that $d_1'' = 6$. Then

$$\log_{10} \frac{d_2''}{d_1''} = \log_{10} 8 ;$$

or, the logarithm of 8 to the base 10, which, by reference to logarithm tables, is 0.9030900. In other words $10^{0.9030900} = 8$.

But we may arrive at the same result by taking the logarithm of 48, or 1.6812413 and the logarithm of 6 = 0.7781513, and subtracting them. We then have

$$x = 10^{(1.6812413 - 0.7781513)}$$
$$= 10^{(0.9030000)}$$
$$= 8.$$

Logarithms are also used to perform conveniently and quickly *involution* or *evolution; i. e.*, to obtain powers, or to extract roots. Thus it would be a troublesome operation to obtain the 12th power of say 15, or to solve the equation

$$x = 15^{12},$$

but with the aid of logarithms this is very simply performed, because the logarithm of 15, is found to be 1.1760913.

$$\therefore 15 = 10^{(1.1760913)}$$

and

$$15^{12} = \left(10^{1.1760913}\right)^{12}$$
$$= 10^{(1.1760913) \times 12}$$
$$= 10^{14.1130956}$$
$$= 10^{14} \times 10^{0.1130956}$$
$$= 10^{14} \times 1.29746$$

since by reference to tables the number

1.29746 has the logarithm 0.1130956, or
$x = 129,746,000,000,000$, so far as six
places of figures. The actual number is
129,746,337,890,625.
Similarly, the 4th root of 15, is obtained
by dividing the logarithm of 15 by 4.
Thus

$$15 = 10^{1.1760913}$$
$$\sqrt[4]{15} = 15^{\frac{1}{4}} = 10^{\frac{1.1760913}{4}} = 10^{0.2910228}$$
$$= 1.968, \text{ approximately, by reference}$$
to tables.

Consequently, $(1.968)^4 = 15$ approximately.

For some calculations the base 10, is in-
convenient, and a base is then adopted
which is more natural. In the theory of
numbers and their exponents, this base, as
far as five decimal places, is the number
2.71828. . . . . and is called the *Naperian
base*, and is usually represented by the
symbol ϵ. A logarithm of this base
is usually called a *natural logarithm*, a
Naperian logarithm, or a *hyperbolic loga-*

rithm, to distinguish it from the common logarithm to the base 10. It is written $\log_e x$. Thus, if $(2.71828\ldots)^n = x$ then $n = \log_e x$.

It can be readily shown that the Naperian logarithm of a number is greater than the common logarithm of that number in a fixed ratio which is, approximately, 2.3026, so that, if we multiply the common logarithm of a number by 2.3026, we obtain its approximate Naperian logarithm.

Thus the common logarithm of 15, is 1.1760913 or $15 = 10^{1.1760913}$.

The Naperian logarithm of 15, is, therefore, approximately,

$$2.3026 \times 1.1760913 = 2.708, \text{ approximately,}$$

or,

$$15 = e^{2.708}, \text{ approximately,}$$

$$= (2.71828)^{2.708}.$$

Thus on page 157, of Thompson's

"Dynamo-Electric Machinery," there occurs the expression:

$$\frac{a}{\pi} \text{ hyp. log } \frac{d_2}{d_1}.$$

This means $\frac{a}{\pi} \times 2.3026 \log_{10} \frac{d_2}{d_1}$ approximately.

CHAPTER VI.

TRIGONOMETRY is the science which deals with angles and their relations in geometrical figures. There are two ways of measuring angles in general use.

The first consists in the ordinary method of dividing a complete revolution into 360°, and measuring the angle in degrees, minutes, and seconds; there being 60 minutes in a degree and 60 seconds in a minute.

The second method, which is important in theoretical treatment as distinguished from practical treatment, measures an angle by the ratio of its arc to its radius. Thus, in Fig. 1, the ratio of the length a, of the arc of the angle α or AOB to the length of the radius r, of the circle on which it is drawn, is called the *radian measure* of the angle α. It is obvious that if the arc α, is the same length as the radius r, the ratio $\dfrac{\alpha}{r}$ will be unity, and this will be a unit angle in radian measure. Such an angle, which is called a *radian*,

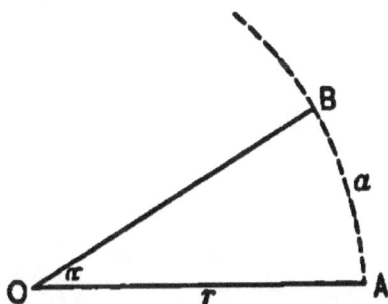

Fig. 1

when expressed in degrees, is equal to 57° 17′ 45″, approximately. A complete circumference, having a length which is 2π times, or 6.2832 times, the length of the radius, such a complete revolution of 360° is equal to 2π radians. Consequently, a right angle is $\dfrac{2\pi}{4,}$ or $\dfrac{\pi}{2,}$ radians, and a single degree is $\dfrac{2\pi}{360} = \dfrac{\pi}{180}$ radians.

In dealing with angles, certain ratios
called the *trigonometrical ratios* or *functions*
are constantly used, and it is important to
clearly understand their nature and mean·
ing. Thus, let *α*, (Fig. 2) be an angle *α*
= *AOB*, included between the lines *OA*
and *OB*. Then let fall a perpendicular
BC, from the point *B*, upon the base *OA*.
Then the fraction whose numerator is the
length of the perpendicular *BC*, and whose
denominator is the length of the radius,
or the fraction $\dfrac{BC}{OB}$ is called the *sine* of the

angle *α*, and is written *sin α*, as an abbre·
viation of the term *sine of angle α*.

Suppose, for example, that the length of
the line *BC*, is 1 inch, while the length
of the line *OA*, which is also the length
of the line, or radius *OB*, is unity, or 1¼
inches. Then the fraction $\dfrac{BC}{OC} = \dfrac{1}{1\frac{1}{4}} = 0.8$

is the sine of the angle *α*; or, in this case,
sin *α* = 0.8.

The fraction whose numerator is the

length between O and C, or the base OC, and whose denominator is OB; that is the fraction $\dfrac{OC}{OB}$, is called the *cosine* of the

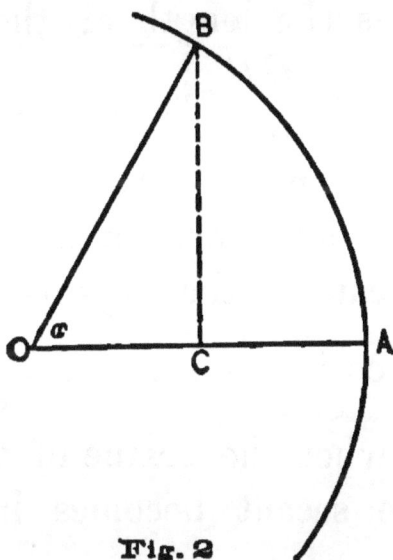

Fig. 2

angle α, and is written *cos* α, as an abbreviation for the term *cosine of angle* α,

The fraction whose numerator is the length of the perpendicular BC, and whose denominator is the length of the base OC, or the fraction $\dfrac{BC}{OC}$, is called the *tangent* of the angle α, and is written *tan* α, which is an abbreviation of the term, *tangent of angle* α.

The fraction whose numerator is the length of the radius OB, and whose denominator is the length of the base OC, or the fraction $\dfrac{OB}{OC}$, is called the *secant* of the angle α, and is written *sec* α, as an abbreviation for the term, *secant of angle* α. The secant of the angle is the reciprocal of the cosine or sec $\alpha = \dfrac{1}{\cos \alpha}$. Consequently, when the cosine of the angle is known the secant becomes immediately known.

The fraction whose numerator is the length of the radius OB, and whose denominator is the length of the perpendicular BC, or the fraction $\dfrac{OB}{BC}$, is called the *cosecant* of the angle α, and is written *cosec* α, as an abbreviation for the term, *cosecant of angle* α. The cosecant of an angle is the reciprocal of the sine of the angle; or, cosec $\alpha = \dfrac{1}{\sin \alpha}$.

The fraction whose numerator is the length of the base OC, and whose denominator is the length of the perpendicular BC, or the fraction $\dfrac{OC}{BC}$, is called the *cotangent* of the angle α, and is written *cot* α, as an abbreviation for the term, *cotangent of angle* α. It is the reciprocal of the tangent of the angle, so that cot $\alpha = \dfrac{1}{\tan \alpha}$.

As an angle increases in value, that is to say, as the radius OB, is carried further and further away from the initial line OA, these trigonometrical ratios—*i. e.*, the sine, cosine, tangent, and their reciprocals, the cosecant, secant, and cotangent—undergo variation. Confining our attention to the *first right angle* or *first quadrant; i. e.*, to the angle α, whose value is not greater than 90°, the sine increases from 0 to 1; the cosine diminishes from 1

to 0; and the tangent increases from 0 to
∞. In other words the sine of 90° is
unity; the cosine of 90° is zero; and the
tangent of 90°, is indefinitely great, or
infinity.

In Fig. 3, the radius at the angle 60°,
has a perpendicular BC_1; assuming that
the radius OA, or OB, is of unit length,
the length BC_1, will be found to be 0.866,
approximately; and this is the ratio of

$$\frac{BC_1}{OB_1}$$

or the sine of the angle 60°; or, sin
60° = 0.866. The length of the base OC,
will be found by measurement to be half
OB; or, if OB, is unit length its value will
be 0.5, so that the cosine of the angle 60°
is 0.5 ; or, cos 60° = 0.5. Similarly, the
fraction represented by the

$$\frac{BC_1}{OC_1} = \frac{0.866}{0.5} = 1.732$$

is the tangent of the angle α; so that
tan 60° = 1.732.

Carrying the moving radius, or *radius vector* as it is called, past the perpendicular *Ob*, into the *second quadrant*, to a position

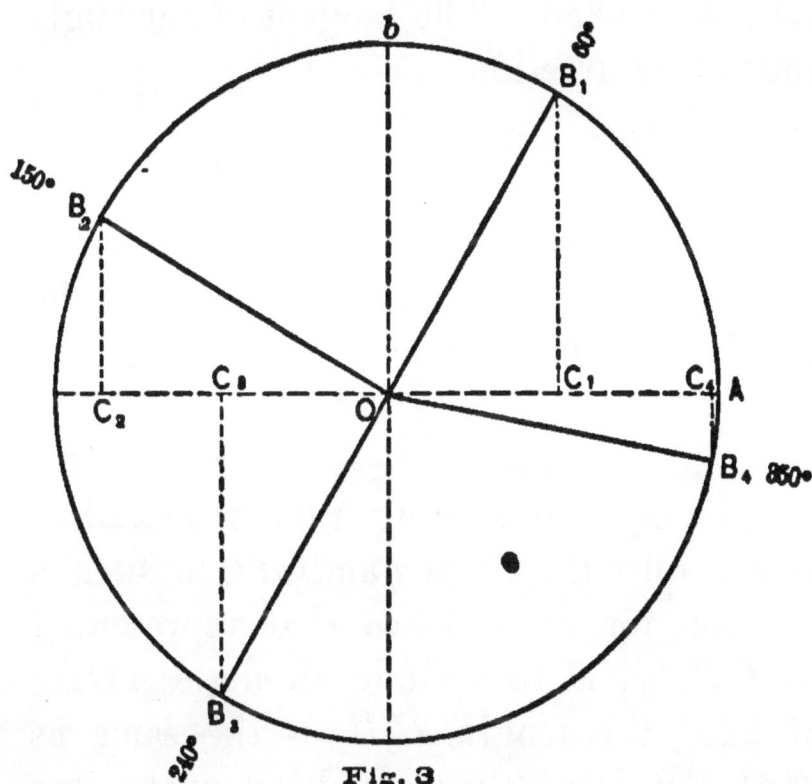

Fig. 3

OB_2, which is 150° angularly distant from OA, the length of the line B_2C_2, which represents the sine of the angle B_2OA, is 0.5, so that sin 150° = 0.5. The length of

the line OC_2' which measures the cosine of
the angle B_2OA, is 0.866 but being measured backward from O, or in the negative
direction, it is written -0.866, so that cos
$150° = -0.866$. The tangent of the angle
$150°$ is the fraction

$$\frac{B_2C_2}{OC_2} = \frac{0.5}{-0.866} = -0.577,$$

approximately, so that $\tan 150° = -0.577$,
approximately. The cotangent would be
the reciprocal of this or

$$\frac{1}{-0.577} = -1.732.$$

Carrying the moving radius or radius
vector into the *third quadrant*, to such a
position, for example, as that represented
by OB_3, so as to include an angle AOB_3,
of 240°, the length C_3B_2, is the same as
B_2C_1, but is now measured below the line
OA, or negatively; so that sin 240° =
-0.866. The cosine $OC_3 = -0.5$ so that
cos 240 = -0.5. The tangent

$$\frac{B_3C_3}{OC_3} = \frac{-0.866}{-0.5} = 1.732,$$

while the cosine, secant, and cotangent are the reciprocals of these three quantities respectively.

Carrying the radius vector into the *fourth quadrant*, to such a position as is represented by the line OB_4, so as to include an angle of 350°, between OA and OB_4, the length B_4C_4', which measures the sine of the angle, will be found to be −0.174, approximately, the minus sign being attached, because B_4C_4' is below the line OA; the cosine, or the length OC_4', is +0.985, approximately, so that cos 350° = 0.985. The tangent

$$\frac{B_4C_4}{OC_4} = \frac{-0.174}{0.985} = -0.176,$$

approximately, and the cosecant, secant, and tangent are found as the reciprocals of these three quantities, respectively.

Trigonometrical tables give the numerical values of the sine, cosine, tangent, cosecant, secant, and cotangent for all angles to any reasonable desired degree of

accuracy.   Tables are also commonly given
of the common logarithms of the trigonometrical ratios for convenience in multiplying and dividing them.

On page 648, of Thompson's "Dynamo-
Electric Machinery," occurs the equation:

$$e_1 = E_1 \cos \theta.$$

This means that the E. M. F. induced in
an adjacent coil of wire, at any instant by
the revolving magnet, is equal to the product of a certain maximum E. M. F.,
denoted by the symbol E_1, and the cosine
of the angle which is included between the
position of the magnet, at the instant considered, and the position of maximum
E. M. F.   If the angle $\theta = 0$, $\cos \theta$, will be
found by reference to a table, or by exami-
nation of Figs. 2 and 3, to be unity, and
the equation becomes $e_1 = E_1$; whereas, if
the magnet has turned through an angle of
90°, it cannot at that instant induce any
E. M. F.   We find, correspondingly, that
$\cos 90° = 0$, so that $e_1 = E_1 \times 0 = 0$.

On page 160, of Thompson's "Poly-phase Electric Currents," the equation occurs:

$$\cos \varphi_1 = \frac{r}{I_1}.$$

Here φ_1, is a certain angle; namely, an *angle of lag*, or the angle between an E. M. F. and the current it is supposed to produce. The equation states that the cosine of this angle φ_1 is the quotient obtained by dividing the resistance r ohms, by the impedance I_1 ohms, and if $r =$ say, 3 ohms, and $I_1 = 5$, then

$$\frac{r}{I_1} = \frac{3}{5} = 0.6,$$

and the equation becomes $\cos \varphi_1 = 0.6$. By reference to a table of cosines it will be found that, for this cosine, $\varphi_1 = 58° 8'$, approximately.

On page 563, of Thompson's "Dynamo-Electric Machinery," occurs the following, in connection with Fig. 362 :

$$\sin \varphi = \frac{pL}{\sqrt{R^2 + p^2 L^2}}.$$

This equation may be regarded either from an algebraic or from a geometrical point of view. First, considering it algebraically, the statement contained in the equation is that if we form a fraction whose numerator is the product of the two quantities represented by p and L, and whose denominator is the square root of the sum of two squares,—one being the square of the resistance R, and the other being the square of a quantity pL,—the quotient will be the sine of a certain angle φ; or, regarding the equation geometrically, it is to be observed that in Fig. 362, the length of the line marked *impedance*, is, by a well-known geometrical proposition, equal to the square root of the sum of the squares of the base R, and perpendicular pL, so that this length representing the impedance is

$$\sqrt{R^2 + p^2 L^2}\,.$$

But in the triangle it is evident that, by definition, the sine of the angle φ, is the quotient of the length pL, by the length of the radius, or the impedance, so that

$$\sin \varphi = \frac{pL}{\sqrt{R^2 + p^2 L^2}}.$$

Similar reasoning applies immediately to the two equations to be found on page 563, which express the cosine and the tangent of the angle respectively.

On page 182, of Thompson's "Polyphase Electric Currents," appears the equation :

$$c_2 = \frac{E}{2R} \sin (pt - 60°).$$

This equation is equivalent to the statement that a certain alternating current c_2, is equal to a fraction, the numerator of which is the product of a maximum E. M. F., E, and the sine of a certain angle $(pt - 60°)$, while the denominator is double the resistance R. The angle $(pt - 60°)$, consists of two terms, the first of which pt, is the product of a constant quantity p, and the time t, in seconds, starting from some assigned moment. As time goes on, the angle pt, continually increases. This angle must be reduced by 60°, and the sine of the remainder, as found in trigonometrical tables, inserted in the equation, will give the value of the current c_2. It will be found, on studying the behavior of this angle $(pt - 60°)$, that it steadily increases from 0, round to 360°, then again round to 720°, and so on, repeating itself in each revolution. The sine of the angle will pass through the values $0 + 1$, $0 - 1$, $0 + 1$, $0 - 1$, and so on, so that the current c_2, will rise to a certain maximum, then diminish to zero, then rise to a negative maximum, and then return to zero cyclically. This is a property which denotes an alternating current. Since the current varies with the sine of the angle $(pt - 60°)$, it may be said to be a sinusoidal current.

On page 159, of Thompson's "Polyphase Electric Currents," appears the following equation :

$$N = A \cos aB_o \sin 2\pi nt.$$

This equation states that the total magnetic flux N, in C. G. S. lines, enclosed by
a certain conducting loop under consideration, is equal to the product of four terms.
The first term is the area A, of the loop expressed in square centimetres. The second
term is the cosine of the angle included
between the plane of the conducting loop
and the plane perpendicular to the direction of the magnetic flux. The third term
is B_0, the *flux density*, or the number of
magnetic lines per normal square centimetre, and the fourth term is the sine of
the angle which is the product of 2 into π,
into n, into t; π, being 3.1416; n, being the
frequency of alternation, and t, being the
time expressed in seconds from a given
epoch. If we form the product $2\pi n t$,
and substitute the sine of this angle for
any given instant t, and also if we find the
cosine of the angle a, which the conductor
occupies at the instant t, the equation will
enable us to determine the total quantity
of magnetic flux embraced by the conductor at the instant t.

CHAPTER VII.

DIFFERENTIAL CALCULUS.

THE *differential calculus* supplies a method of dealing with quantities that are subjected to continued variation and of which the rate of variation at any instant is required.

Imagine a railway train, in operation upon a track, and suppose that it is necessary, for any purpose, to know the speed with which the train moves; or, in other words, to determine the velocity of the train. This velocity might be determined experimentally by some form of centrifugal indicator, properly calibrated, but in the absence of such an instrument it might be necessary to determine the velocity by ascertaining how far the train moved along the track in a given measured interval of time. For example, with the aid of a stop-watch, it might be perfectly feasible to count the number of rail lengths passed by in one minute, each rail being just 30 feet long. If, in this way, we found that in one minute the train passed over exactly 150 rails, the total distance traversed in the minute would be 4,500 feet, and this represents a velocity of 4,500 feet per minute.

This method might be capable of a high degree of accuracy if the train were moving at a uniform speed; but suppose the train is either accelerating, or is being retarded. It is impossible to obtain the correct speed at any moment by an observation lasting for a minute, because, during that time, the train might have greatly varied its speed, and only what might be called an average speed could be obtained by such an observation. If, however, it were feasible to make an observation in one second of time, instead of one minute, it is evident that the degree of accuracy would be greatly increased, because the speed would not have had time to appreciably vary during that interval of one second in which the observation was made. Thus, if in a single second the train moved over exactly 1.5 rails, this would be a distance of 45 feet, and the velocity at the time considered would be 45 feet per second, or 2,700 feet per minute. But for sudden starting or sudden stopping it is conceivable that even a single second of time would be too long an interval to measure the speed correctly, because even in one second the speed might vary considerably. If, however, it were possible to measure the distance passed through in say 1/1000th of a second, the degree of accuracy, in determining the velocity at any particular instant, would be much greater, because, during so short an interval of time as the 1/1000th part of a second, very little change in velocity could take place.

In the same way it is evident that for a

theoretically perfect measurement of the
velocity at any given instant, as distinguished from the average velocity extend-
ing over any period of time, the
measurement of the distance run through
would have to be made in an infinitely
short period of time. For example, if the
distance travelled by the train in one
second were 1.5 rails, or 45 feet, representing a mean velocity during that second of
45 feet per second, yet we might find that in
the particular 1/1000th of a second during
that time the distance travelled through
was 1/20th of a foot. The velocity would
be expressed by the fraction

$$\frac{\text{Distance}}{\text{Time}} = \frac{0.05 \text{ foot}}{0.001 \text{ second}}$$

and the quotient would be 50 feet per
second.

Generalizing this, if we represent by the
symbol Δf, (delta f) a certain small distance in feet, and by Δt, (delta t) a certain

small interval of time during which the
observation lasted, the mean velocity during the time Δt, would be:

$$\text{mean } v = \frac{\Delta f}{\Delta t},$$

The smaller we make Δt, the smaller
will the distance travelled Δf, become,
and the closer will the mean velocity so
computed approach to the instantaneous
velocity at the moment considered. If we
could imagine that an indefinitely small interval of time was selected, that is to say,
if Δt, was indefinitely shortened, the corresponding value of the distance travelled
would be also indefinitely reduced, but the
quotient formed by dividing the infinitely
small distance Δf, by the infinitely small
time Δt, would give the theoretically exact velocity of the train at that moment.
In the symbols used by the differential
calculus, this indefinitely small time is
written dt, and the corresponding inde-

finitely small distance is written df, the velocity being

$$v = \frac{df}{dt}$$

It is to be observed that df, considered alone, or dt, considered alone, becomes meaningless, because we cannot conceive of an infinitely short time, or of the distance the train would run in an infinitely short time, but it is perfectly feasible to imagine that the velocity, which is only a mean velocity when the time is expanded to an actual time Δt, and the corresponding distance df, expanded to an actual distance Δf, becomes closer and closer to the real velocity as Δf and Δt, are shortened, and we merely use the symbol $v = \frac{df}{dt}$ in the above equation as expressing symbolically the following statement:

The true velocity of the train is such that if the measurement could be made in an infinitely brief interval of time, and the corresponding distance run through in that time could be observed, the infinitesimally small distance divided by the infinitesimally small time would be the true or instantaneous velocity, although, in any actual measurement, this value can only be approached.

The fraction $\frac{df}{dt}$, is called the *differential coefficient* of the quantity f, with respect to the quantity t. In the above considered case it is the differential coefficient of the distance travelled by the train to the time of travel. Here the symbol d, cannot be considered as either a product, or as being separable from the quantity to which it is prefixed. But df, is to be considered as an abbreviation for the term, the *differential of f*, and dt, the denominator, is to be considered as an abbreviation for the term, the *differential of t*. The ratio $\frac{df}{dt}$, has usually a definite limiting value, as, in the above case, the true instantaneous velocity of the train, although df and dt, are each assumed to be indefinitely small.

A differential coefficient $\frac{df}{dt}$, is, there-fore, to be regarded as the limit which the fraction $\frac{\Delta f}{\Delta t}$ assumes, when Δt, is indef-initely reduced ; *i. e.*, the limiting ratio of a small change in f, to the small change in t, which gives rise to it.

Almost the whole theory of the differen-tial calculus is devoted to a consideration of the values which differential coefficients assume under different circumstances ; *i. e.*, of the limiting values of the ratios between finite quantities, which are dependent upon each other, or connected together in some definite manner, and which are coincident-ally reduced indefinitely.

On page 721, of Thompson's "Dynamo-Electric Machinery," occurs the following equation, which is called a differential equation because it involves differential coefficients:

$$E_1 - M\frac{dC_2}{dt} - L_1\frac{dC_1}{dt} - R_1 C_1 = 0.$$

Here the equation makes the following statement:

In the primary circuit of an induction coil there are, in general, at any instant, four E. M. Fs. acting, and their sum is zero.

The first term is a certain given E. M. F. E_1, usually called the *impressed E. M. F.*

The second term is $- M\frac{dC_2}{dt}$, and is the product, taken negatively, of the mutual inductance M, between the primary and secondary windings of the coil, multiplied by the differential coefficient of the secondary current C_2, with respect to time. We can imagine that by some process the exact rate of increase of the secondary current is determined by observing the actual increase ΔC_2, which takes place in an very short interval of time Δt, and the ratio $\dfrac{dC_2'}{dt}$, when the interval of time Δt is reduced theoretically to zero, is the true instantaneous rate of increase of the secondary current at the instant of time considered.

The third term is the product, taken negatively, of the self-inductance L, of the primary coil, multiplied by the instantaneous rate of increase of the primary current, expressed by the differential coefficient $\dfrac{dC_1'}{dt}$. Here the increase in primary current dC_1' is supposed to be measured in an indefinitely brief period dt, of time, and the ratio or quotient obtained by dividing the imaginary change of primary current, so measured, by the infinitesimal interval dt, in which the change takes place, is the instantaneous true increase of primary current at that instant.

The fourth term is the product, taken negatively, of the drop of pressure, or

E. M. F., in the circuit, due to the ohmic resistance, consisting of the product of the primary resistance and the primary current.

The equation, then, states that the sum of all the E. M. Fs., including the apparent E. M. F. of drop, is equal to zero in the primary circuit at any and every instant of time.

There are rules for determining the differential coefficient of any variable, in terms of any other variable upon which it depends. Thus, if a certain variable is expressed by the equation

$$f = \frac{gt^2}{2}$$

so that f, is a quantity, say a distance travelled, which varies with the quantity t, say the time, in the manner described by the equation, it is shown, in treatises on the differential calculus, that the differential coefficient $\frac{df}{dt}$, is in this case expressed as gt; or,

$$\frac{df}{dt} = gt \text{ when } f = \frac{gt^2}{2}.$$

In reading technical works, however, it is
usually quite unnecessary for the student
to understand how the differential coefficients are obtained, and the reasoning is
entirely unaffected, if the conception of a
differential coefficient is kept in mind.

A differential coefficient $\dfrac{df}{dt}$, where f,
represents distance or space, and t, represents time, is an instantaneous time-rate of
the increase in f.

A differential coefficient $\dfrac{dW}{ds}$, where W,
is the work done, and s, is the space traversed, is an instantaneous space-rate of
increase of work, and is the limit of a
definite amount of work W, performed in
a definite space s, when the space is indefinitely diminished.

On page 159, of Thompson's "Polyphase Electric Currents," occurs the equation:

$$E = -\frac{dN}{dt}.$$

This expresses the fact that the E. M. F.
induced in a conducting loop is the negative instantaneous time-rate of increase of
magnetic flux through the loop, or the
limit of the quantity of flux added to the
loop in unit time, when the unit of time is
made indefinitely small; in other words,
an instantaneous velocity of adding flux to
the loop.

Immediately above this equation, on
page 159, is the solution of this differential
coefficient for the case assumed, in which
$N = A \cos aB_o \sin 2\pi nt$. This the student will have to accept without demon-
stration until he has so far familiarized
himself with the processes of differentia-
tion, according to the rules of differential
calculus, as to be able to perform the
solution. But for the purpose of following
the explanations in the text on page 159,
this knowledge is quite unnecessary.

CHAPTER VIII.

JUST as the operation of extracting a root is, as we have seen, the inverse operation of raising a quantity to the corresponding power, so the integral calculus deals with the inverse operation to that of the differential calculus. In the differential calculus one of the principal objects is to determine, from a known relation between two connected variables, such, for example, as elevation and barometric pressure, what is the instantaneous rate at which, at a given elevation, the pressure varies as the elevation is changed; or, the ratio of an extremely small change in pressure to the extremely small change of elevation that produces it, this ratio being called the *differential coefficient* of the pressure with respect to elevation. Inversely, one of the principal objects of the integral calculus is to determine, from a given differential coefficient; *i. e.*, in the above case, from a known instantaneous rate of increase of pressure with elevation, what must be the law connecting the variables; *i. e.*, the general relation between pressure and elevation.

The problem of the integral calculus, or of *integration*, is, therefore, that of finding the relations which give rise to a differential coefficient. It can be shown to resolve itself into a summation of an indefinitely long series of terms, each of which is indefinitely small.

Suppose, for example, we allow a stone to fall to the ground from an elevation. It is known that, neglecting the friction of the air, the velocity of the stone at any moment is proportional to the time which elapses from its release. If this velocity be graphically represented as shown in Fig. 4, by distances above the line *OT*, along

which time is measured in seconds, then
the line $O V$, will show the velocity at any
moment.   Thus, at an interval of one second after the release of the stone, the
velocity acquired will be 32.2 feet per
second; after two seconds, 64.4 feet per

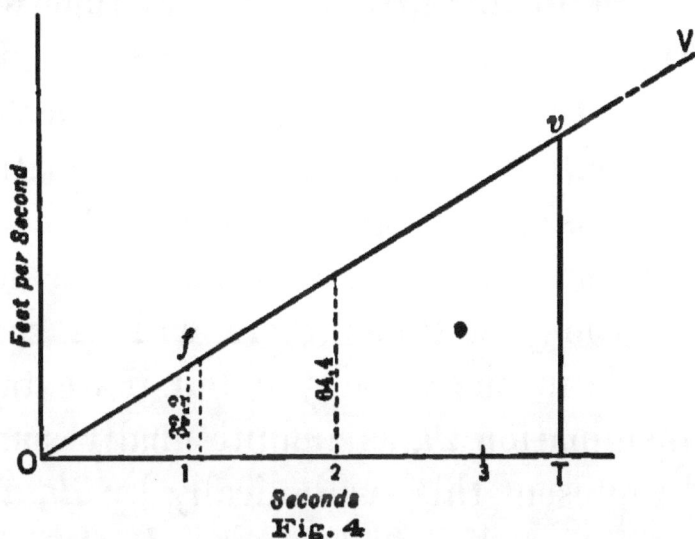

Fig. 4

second; and after t seconds, $t \times$ 32.2 feet
per second.

Suppose it be required to determine the
total space through which the stone falls in
a time T seconds, T, being 3.5 seconds in
Fig. 4.   Since the velocity is varying all
the time, we cannot say, from a mere

inspection of the problem, what the total distance will be, but it is evident, that if we consider any very small interval of time Δt, say at one second from the start, or when the velocity v, is 32.2 feet per second, then the distance passed through by the stone in this little interval of time will be the product of the velocity v, then existing, and the time interval, or, the space Δs, equals $v\Delta t = 32.2\Delta t$. Such an equation as this is never strictly correct, because in the interval of time Δt, no matter how small it may be taken, there will be some variation in the velocity v, but if we take in imagination Δt, as infinitesimally small, and represent this symbolically by dt, and the corresponding distance by ds, then we shall have strictly, $ds = vdt$. Or, in **Fig. 4**, at f, one second after release, will be

$$ds = 32.2dt.$$

This product represents the area of a little strip between the dotted lines connecting 1 and f, in Fig. 4. The smaller dt, is taken, the narrower will be this strip ; and, if dt, is in imagination infinitely small, the strip dt, will be of vanishing width.

If we divide the whole time T, into equal infinitesimal intervals dt, the equation $ds = vdt$, will be true for every little strip drawn as in Fig. 4 ; v, however, varying from strip to strip, being 64.4 at 2 seconds, 96.6 at three seconds, and so on. We have then an imaginary series of equations like the following :

$$ds_1 = v_1 dt = 0 \times dt,$$

because the velocity starts at zero.

$$ds_2 = v_2 dt,$$
$$ds_3 = v_3 dt.$$
$$\cdots\cdots\cdots\cdots$$
$$ds_m = v_m dt = 32.2\ dt;$$

m, being the number of intervals in one second, and 32.2, being the velocity at the end of one second ; until, finally,

$$ds_n = v_n dt;$$

v_n, being the velocity at the end of the interval T. If we add all these equations together, we obtain the total distance fallen through by the stone in the time T; or,

$$S = ds_1 + ds_2 + ds_3 + \ldots + ds_n$$
$$= v_1 dt + v_2 dt + v_3 dt + \ldots + v_n dt.$$

This sum is written in the notation of the integral calculus,

$$S = \int_0^T v\,dt\,;$$

$i.\ e.$, the sum, represented by the sign $\int$, of all the elementary terms whose type $v\,dt$, taken from the term at $t = 0$, and continued up to the term $t = T$, and assuming that the number of terms is indefinitely increased. In Fig. 4, the area of all elementary strips of the type $v\,dt$, will be the area of a triangle oTv, and this will represent the space fallen through by the stone in the time T.

The natural problem of the differential calculus applied to the falling stone would be as follows :

A stone falls through a distance of 16.1 feet in 1 second, and 197.225 feet in 3.5 seconds, at a continuously varying velocity according to some definite law. What is the instantaneous velocity, or the differential coefficient of the space traversed with respect to the time?

The answer would be:

$$V = \frac{ds}{dt} = 32.2t.$$

The natural problem of the integral calculus applied to the stone is just the opposite.

Having given the known relation that the velocity of the stone at any time t, seconds after its release, is

$$V = 32.2t \text{ feet per second,}$$

what is the total distance which it will describe in a given time T?

We know that when $T = 3.5$, $S = 197.225$, and S, is obtained by the summation of a theoretically infinite number of infinitely small terms, and is represented graphically by the area comprised between the line of velocity ov, the perpendicular Tv, and the base oT', this area being composed of an infinite number of little vertical strips side by side and of the type $f dt$.

On page 25, of Thompson's "Polyphase Electric Currents," appears the following formula:

$$\frac{1}{\psi} \int_0^\psi e.\cos \gamma. \, d \gamma.$$

This expression means that a certain integral is to be divided by the quantity ψ; or, what is the same thing, multiplied by the reciprocal of ψ. The integral is the sum of an indefinite number of elements, each of which is of the type $e \cos \gamma \, d \gamma$, and which elements must be summed up between the limits of $\gamma = 0$ and $\gamma = \psi$.

In Fig. 5, let the line oA, be marked off to correspond to the angle γ. For example, each inch of the line oA, might correspond to some definite number of degrees or radians of the angle γ, considered, any suitable scale being chosen. Let distances measured vertically above this line oA,

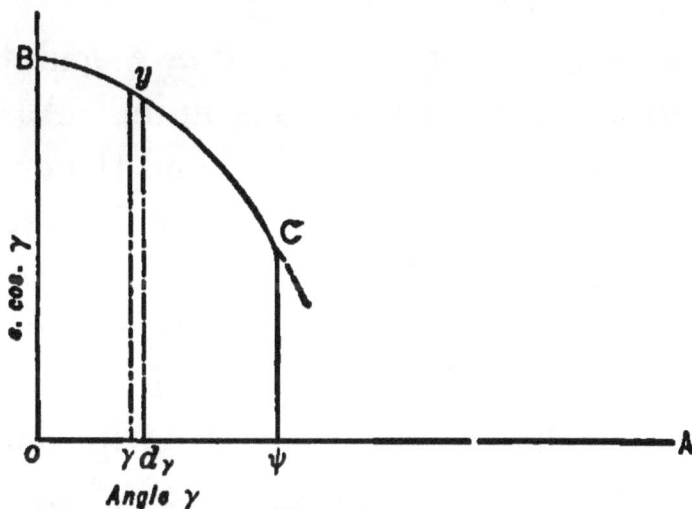

Fig. 6

and, therefore, drawn parallel to oB, represent the value of the product $e \cos \gamma$.

As γ, increases, $\cos \gamma$, will vary. Consequently, the product of e and $\cos \gamma$, will vary, and the elevation which represents $e \cos \gamma$, will vary. The curve BC, is supposed to be correctly drawn, so that at any point y, the perpendicular γy, corresponds in length to the value of $e \cos \gamma$, for the particular angle represented by the distance of the foot of the perpendicular γ, from o, so that length of γy, equals $e \cos \gamma$. Similarly, length oB, equals $e \cos o$, and length ψC, equals $e \cos \psi$.

Since $\cos 0 = 1$, $e \cos 0 = e$, and the length of oB, is equal to e, units. As γ, increases $\cos \gamma$, diminishes and the curve ByC, falls to the minimum C, which represents $e \cos \psi$. At γ, if we take a small length on the base equal to $d\gamma$, and carry perpendiculars from the ends of this small space, so that we have a thin strip shown by the dotted lines, the area of which is $e \cos \gamma \, d\gamma$, the height of such a strip will, of course, vary on different points along the base; but the sum of all such strips, taken between the limits of the base $\gamma = 0$ and $\gamma = \psi$, will be the total area between the line oB, the curve BC, the line ψC, and the base $o\psi$, and if we suppose $d\gamma$, to be indefinitely small, so that the strip becomes indefinitely narrow, the sum of all such strips will give accurately the included area $BC\psi o$. The integral,

$$\int_0^\psi e \cos \gamma \, d\gamma \,.$$

will be numerically equal to the area $BC\psi o$, since it will be the sum of all the strips whose height is $e \cos \gamma$, and whose width is $d\gamma$, the latter being taken indefinitely small and the terms being taken in correspondingly great number.

A problem expressed by an integral may, therefore, always be considered as equivalent to the summation of an area between a specified curve, conforming to the given law, and the base line, between definite perpendiculars on the base.

The rules for performing integration must be acquired by practice and acquaintance with the integral calculus. In most technical books it is sufficient for the purpose of the reader if the conception of an integral is clearly apprehended.

INDEX.

A

INDEX.

F

G

I

L

S

T

V

www.ingramcontent.com/pod-product-compliance
Lightning Source LLC
Chambersburg PA
CBHW021828190326
41518CB00007B/776